Grade 2

Science Resources

Developed at
The Lawrence Hall of Science,
University of California, Berkeley
Published and distributed by
Delta Education,
a member of the School Specialty Family

© 2014 by The Regents of the University of California. All rights reserved. No part of this book may be reproduced or transmitted in any form or by any means, electronic or mechanical, including photocopying or recording, or by any information storage and retrieval system, without permission in writing from the publisher.

1361722
978-1-60902-492-5
Printing 2 — 1/2014
Courier, Kendallville, IN

Table of Contents

Physical Science
Balance and Motion

Make It Balance! **9**
Push or Pull? **19**
Things That Spin **24**
Rolling, Rolling, Rolling! **29**
Strings in Motion **38**

Table of Contents

Earth Science

Air, Weather, and Earth

Matter **47**
What Is All around Us? **51**
Clouds **58**
What Is the Weather Today? **60**
Water in the Air **70**
Where Is Water Found? **77**
Understanding the Weather **87**
Changes in the Sky **93**
Seasons **106**
Getting through Winter **114**
Exploring Rocks **122**
Comparing Solids and Liquids **131**
Heating and Cooling **141**
Resources **147**

Table of Contents

Life Science

Insects and Plants

Animals and Plants in Their Habitats **167**
Flowers and Seeds **183**
So Many Kinds, So Many Places **192**
Variation **198**
Insect Shapes and Colors **205**
Insect Life Cycles **212**
Life Goes Around **221**
Fossils **235**

Table of Contents

References

Science Safety Rules **248**
Outdoor Safety Rules **249**
Tools for Scientific Investigation **250**
Glossary **266**
Photo Credits **272**

Physical Science
Balance and Motion

Table of Contents

Make It Balance! **9**
Push or Pull? **19**
Things That Spin. **24**
Rolling, Rolling, Rolling! **29**
Strings in Motion **38**

Make It Balance!

We live in a world full of **motion**. But not everything moves in the same way.

Some things move from one place to another. Some things **spin** around and around.

Other things **balance**. They might move only if you give them a little **push**.

At the circus, you can see people balancing in amazing ways. They might balance on ropes or on chairs. It takes lots of practice.

With some practice, you can balance on a log. Is it easier with your arms down by your sides or with your arms out?

Try balancing on one foot. Is it easier with your eyes open or closed?

In some countries, people balance things on their heads. This is how they carry things from place to place.

What can you balance on your head? Try a book or an apple. How far can you walk and keep it balanced?

What will happen to this toy if you gently push on it? It might wobble. But it will return to where it started. It will return to a **stable position**. This toy is **counterbalanced** to help it come back to balance, even if it is pushed.

You see things balancing all around you. Which pictures show things that are balancing?

What keeps things in a stable position?

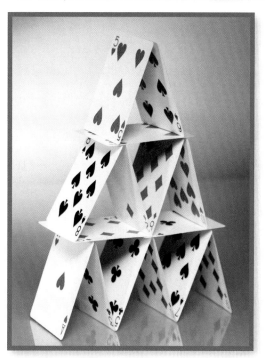

Review Questions

1. Think about balancing on one foot. What can you do with your body to help you balance?

2. Is it easier to balance a ball or a book on your head? Why?

3. What does *balance* mean?

4. What does *counterbalance* mean?

Push or Pull?

One way to make things move is to push them. Another way to make things move is to **pull** them.

A push or a pull is called a **force**. You always need a force to make something happen. A push or a pull can make something move, stop it from moving, or change its direction.

You can push things with your hands or body.
Moving **air** can also push things.

When you play baseball, do you use the bat to push or pull the ball?

It doesn't matter if you hit a home run or strike out. **Gravity** always pulls the ball to the ground. Gravity is a pulling force.

Think about a roller coaster. What pulls it up the first hill? What force pulls it back down again?

Push or pull, a force is always needed to make things move.

Review Questions

1. What are two things a force can do to an object's motion?

2. Tell about one way you can move a ball. Is the force a push or a pull?

3. Tell how to spin a pinwheel, and describe the force.

4. Think about throwing a ball into the air. What will gravity do to this ball?

Things That Spin

Things that spin are all around us. When something spins, it turns on its **axis**.

Tops need to spin fast to balance. What happens when a top slows down?

Some things spin slowly, like this Ferris wheel.

Look at these pictures.

Which things spin fast?
Which things spin slowly?
Which things don't spin at all?

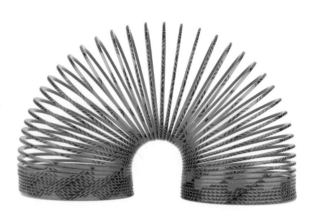

Review Questions

1. What does *spin* mean?

2. Name four things that spin.

3. Name three things that don't spin.

4. How does a top stay balanced?

Rolling, Rolling, Rolling!

Things can move by rolling. When something **rolls**, it goes around and around. But instead of staying in one spot, it moves from one place to another.

Things that have round surfaces roll easily. Marbles roll, and so do cans.

Look at the wheels on this page. Are the wheels spinning or rolling? What's the difference between spinning and rolling?

Some things don't roll in a straight line. Things that are shaped like a paper cup don't roll straight.

Some things don't roll at all. Things that are flat won't roll.

Which of these things will roll down the ramp? Which things will slide?

Gravity helps things roll downhill. It's fun to see how fast something will go.

Which ramp would you use? Why?

Ready, set, go!

Things move in different ways. We can look for patterns in how things move. Spinning things go around and around. Rolling things go around and ahead.

This toy bear can roll and balance. Can you see why?

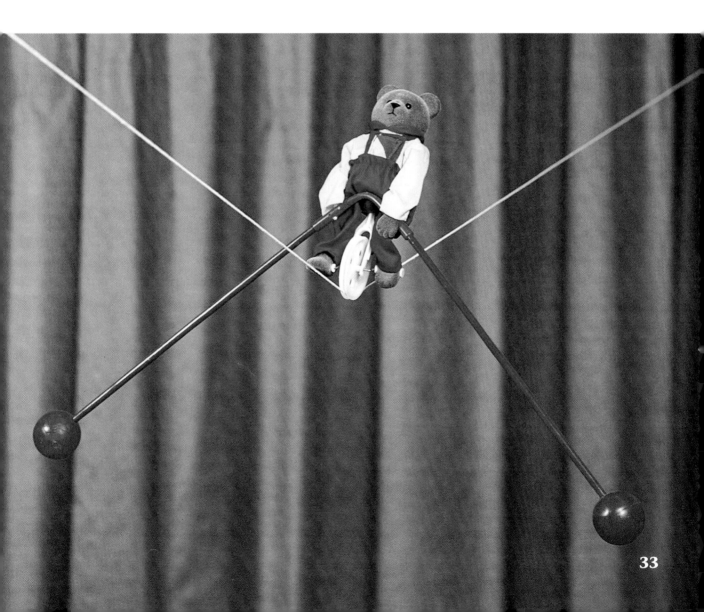

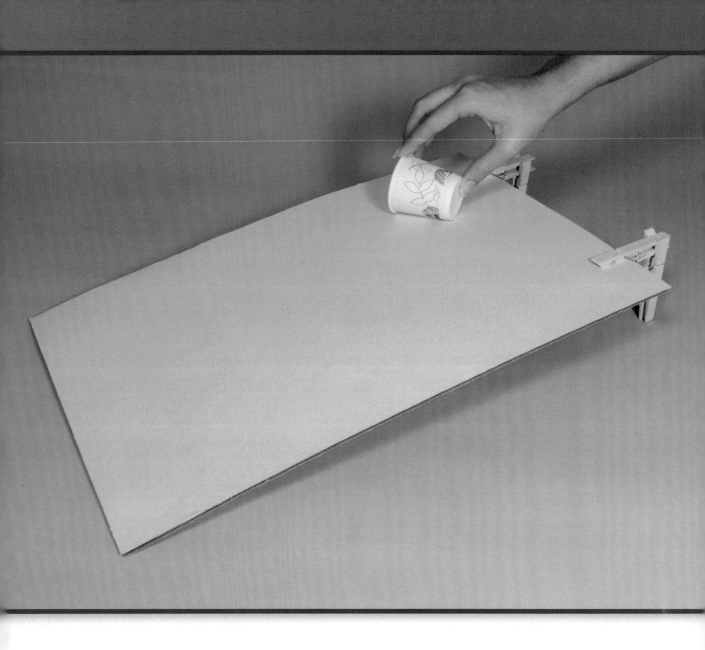

To find an object, we need to know its **position**. Position tells us where something is. What is the position of this cup on the ramp?

Motion is a change in position. How can you tell if an object is moving? See if the object's position changes. The cup rolled off the ramp. Can you describe the change in its position? Can you show how the cup moved?

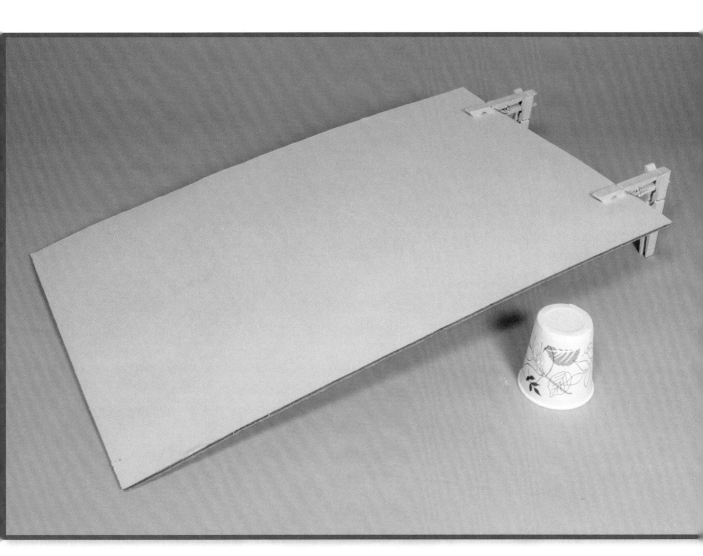

Fast, slow, up, down, spinning, rolling, sliding . . .
Almost everything moves!

What ways can you move?

Review Questions

1. Name four things that roll.

2. Describe what happens to a tennis ball when it rolls on the floor.

3. Compare rolling and spinning. How are they alike? How are they different?

4. Compare rolling and sliding. How are they alike? How are they different?

Strings in Motion

Some musical instruments have strings. Guitars, harps, and violins are stringed instruments. The strings move, and you hear **sounds**. How do moving strings make sounds?

A harp's strings move back and forth when you pluck them. Back-and-forth motion is called **vibration**. Vibrating strings push on the air. This motion makes sound waves. The waves move out in all directions. When sound waves enter your ears, you hear sound.

Short strings vibrate faster than long strings. Strings that vibrate fast make high sounds. They sound like ting, ping, bing.

Long strings vibrate more slowly than short strings. Strings that vibrate slowly make low sounds. They sound like boom, thum, dom.

A harp can make very soft sounds, like a whisper. Or it can make big, loud sounds, like a yell. Soft and loud are **properties** of sound called **volume**. Loud sounds have more **energy** than softer sounds do. A harp player changes the volume of the sound by how hard she plucks each string.

A harp player chooses the **pitch** by plucking strings of different lengths. When the strings move in just the right way, you hear beautiful music.

Which one of these stringed instruments makes high-pitched sounds? Which one makes low-pitched sounds?

Review Questions

1. What happens when you pluck strings on a harp?

2. What do you call the back-and-forth motion of strings?

3. Would you pluck a short string or a long string to make a low-pitched sound?

4. Do long strings vibrate fast or slowly?

Earth Science
Air, Weather, and Earth

Table of Contents

Matter	47
What Is All around Us?	51
Clouds	58
What Is the Weather Today?	60
Water in the Air	70
Where Is Water Found?	77
Understanding the Weather	87
Changes in the Sky	93
Seasons	106
Getting through Winter	114
Exploring Rocks	122
Comparing Solids and Liquids	131
Heating and Cooling	141
Resources	147

Matter

The world is made up of many things. Mountains, trees, **air**, and water are just some of them. These things may all seem very different. But in one way, they are all the same. They are all **matter**. Matter is anything that takes up space.

Matter can be divided into three groups called **states**. They are **solid**, **liquid**, and **gas**.

A bridge is a solid.

Water is a liquid.

Bubbles are filled with air. Air is a gas. How did the gas get into the bubbles?

Review Questions

1. What is matter?

2. What are the three states of matter?

What Is All around Us?

We can't see it, but it's all around. It's in the sky. It's in the treetops. It's on the ground. It's near and far, high and low. What is it?

Air! We cannot see air, but we know it is there. What happens when you blow up a balloon? You fill it with air. You can see that air takes up space.

You also can feel air on your skin when the **wind** blows.
Wind is moving air.

We can tell that air is there when we fly a kite. The wind **pushes** against the kite and keeps it in the sky.

We can tell that air is there when a parachute floats to the ground. Air pushes up against the parachute so it comes down slowly.

Even this boat shows us that air is all around. A propeller on the back of the boat pushes on the air. The boat moves forward.

So, what is all around us, everywhere we go?
You know!

56

Review Questions

1. Where is air?

2. What is wind?

3. How do you know air is there?

Clouds

Clouds can be many sizes and shapes. Watching clouds can help tell you what the **weather** will be.

Some clouds are high in the sky. They are thin and white. These clouds are called cirrus clouds. Cirrus clouds usually mean fair weather.

Some clouds are not so high. They are big, white, and fluffy. These clouds are called cumulus clouds. Cumulus clouds can mean fair or stormy weather.

Some clouds are low to the ground. They are gray and long. These clouds are called stratus clouds. Stratus clouds usually mean **rain**.

Can you find each type of cloud on the next page?

What Is the Weather Today?

Lots of things happen in the air. The **temperature** might change from warm to cold. Clouds might form, and rain might fall. The wind might start to blow. The condition of the air outdoors is called weather.

How do you know what the weather will be today? One way to find out is to look outside. If the sky is dark and cloudy, you know it might rain. If there are no clouds and the **Sun** is shining, you know it won't rain. Well, it won't rain right away.

Clouds are made of little drops of water. If there is a lot of water in a cloud, the cloud looks gray. The water drops might get bigger and bigger. They might fall as rain.

If the air is cold enough, the water drops might fall as **snow**.

Some days the clouds seem to sit on Earth's ground and water instead of floating in the sky. The air feels wet, and you can't see very far. These clouds near the ground and water are called fog.

The weather you see when you look outside might change. A day might start out bright and sunny.

Later, clouds might form. Soon, the sky is filled with them.

If the clouds hold enough water and the drops get big enough, it will rain. A **storm** can blow in and out in an afternoon. Or, a storm can stay around for days.

Weather is in the air. Air is all around you. You feel and see the weather every day, all the time. So, go outside. Enjoy the weather. It might change tomorrow!

Review Questions

1. What is weather?

2. What weather have you observed?

3. Think about the weather today. Is the air hot or cold? Is it windy? Rainy? Cloudy?

Water in the Air

Sometimes it rains. Water falls from the clouds.
The water flows down sidewalks and streets.
It forms puddles in low spots.

Then, it stops raining. The heat from the Sun warms the surface of the land. The heat from the Sun also warms the air and water. Soon, the puddles of water are gone. The sidewalks and streets are dry. Where does the water go?

The water dries up. The water changes from a liquid to a gas. Water as a gas is called **water vapor**. The water vapor goes into the air. This is called **evaporation**. We can't see water vapor in the air. Water vapor is invisible.

Have you seen drops of water on the outside of a cold glass of milk? Where do those drops come from?

Those drops of water come from water vapor in the air. This is called **condensation**. When the water vapor hits a cold surface, it changes into liquid water.

Water vapor is all around us in the air. The air moves from place to place as wind.

When the water vapor cools, it condenses. It changes to tiny drops of liquid water. Clouds form. Clouds are lots of liquid water drops. Rain comes from the liquid water in clouds.

Rain falls from clouds as drops of water. Evaporation is liquid water going into the air as water vapor. Condensation is water vapor forming drops of liquid water. **Precipitation** is rain falling from clouds. This pattern is called the **water cycle**.

Review Questions

1. What happens to puddles when the Sun warms the ground?

2. How do clouds form?

3. What do we call rain falling from clouds?

4. What is the water cycle?

Where Is Water Found?

Water is found everywhere on Earth. Water is part of every living thing. Every plant and animal is made of water. Even you are mostly made of water!

Fresh water is found in streams and rivers. Streams can be small like a creek. Rivers are larger streams of water.

Streams and rivers have flowing water. The water from streams and rivers flows into and out of ponds and lakes.

Fresh water is found in ponds and lakes, too. Ponds are small bodies of still water. Lakes are larger and deeper bodies of water.

The water moves slowly in ponds and lakes. Sand and silt settle to the bottom of ponds and lakes.

Fresh water is our most important **natural resource**. Plants and animals need water to live and grow. People use water to drink, cook, and wash. People use water to grow food and to power factories, too.

Most of the water on Earth is **salt water**. Salt water is found in seas and the ocean. The ocean is the largest body of salt water. Seas are smaller than the ocean.

Salt water is found in salt **marshes**. They are muddy places next to seas. Salt marshes have lots of grasses and small plants. Salt marshes have slow-moving water.

Salt water is found in mangrove forests. They are like salt marshes, but they have trees and bushes. The roots of mangrove trees help protect the shore.

Salt water is found in coral reefs. Coral reefs grow in warm, shallow seas. Coral reefs are made from corals. Corals are the hard parts of sea animals.

Salt water is found on sandy beaches and rocky shores, too. You can see the ocean water move back and forth in waves on beaches and shores.

Review Questions

1. How is the water in lakes and rivers different from the water in the ocean?

2. How does the water in ponds and lakes compare to the water in rivers and streams?

3. Look at the pictures above. Compare the properties of water.

4. Where is fresh water found in your community?

5. Where is salt water found in your community?

Understanding the Weather

Some scientists study the weather. They are called **meteorologists**. They use instruments to gather information about the weather. Meteorologists **measure** the temperature of the air. They watch clouds form. They measure how much rain falls. They measure wind speed and direction.

Weather balloons carry weather instruments high into the sky. The weather instruments gather information. This information helps meteorologists tell us what the weather will be.

Sometimes weather is dangerous. Meteorologists can help us know when to get ready for a storm.

A **tornado** is a twirling, cloudy storm. A tornado's winds blow around and around very quickly.

A **hurricane** is a very strong, wet, and windy storm. Hurricanes form over warm ocean water.

A thunderstorm is a storm with lightning. Lightning can be dangerous. It is important to learn safety rules to be prepared for storms.

Review Questions

1. What is a meteorologist?

2. What does a meteorologist do?

3. Why is it important to know the weather?

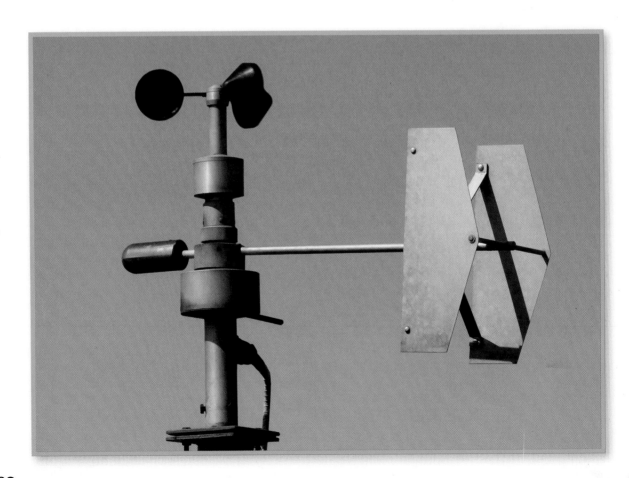

Changes in the Sky

When you look up at the sky, what do you see? It depends on the time of day. It depends on the time of year, too.

Sometimes you see the Sun. The Sun is a **star** close to Earth. You can feel the Sun's warmth and see it shine. When you can see the Sun's light, it is daytime.

Where do you see the Sun in the morning?
It is low in the sky in the east.

Where do you see the Sun just before it gets dark?
It is low in the sky in the west.

Where do you think the Sun is at noon?

Where do you see the Sun at night? You can't see the Sun because it isn't in the night sky. The sky is dark without the Sun in the sky. The Sun makes day and night.

When do you see other stars in the sky? You see other stars only at night. It has to be dark to see them.

Here are stars we see in the summer sky and the winter sky. Do they look like the same pattern of stars?

summer

winter

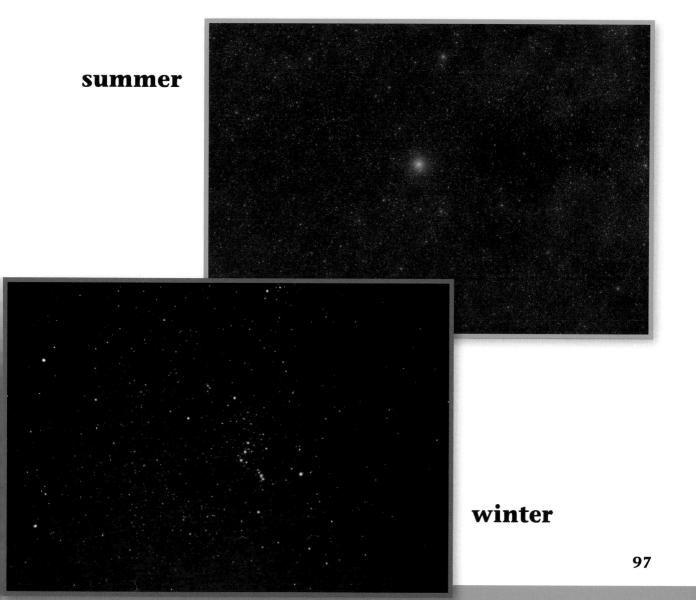

Sometimes you see clouds in the sky. It is easier to see clouds in the day sky. But clouds can be in the night sky, too.

Clouds move with the wind. They change all the time. Sometimes clouds block the Sun. They make **shadows** on the ground.

Sometimes you see the **Moon** in the sky. You can see the Moon in the day sky and the night sky. But it looks different at different times.

This is a full Moon. You can see a full Moon only at night.

Sometimes the Moon looks like a smile. This shape is called a crescent Moon. You can see a crescent Moon in the day sky and the night sky.

Sometimes the Moon looks like a half circle. This shape is called a quarter Moon. You can see a quarter Moon in the day sky and the night sky.

Sometimes the Moon looks like an egg. This shape is called a gibbous Moon. You can see a gibbous Moon in the day sky and the night sky.

Look for the Moon every day or night for a month. Record what you see and when you see it. Is there a pattern to the Moon shapes you see?

Month _____

Sunday	Monday	Tuesday	Wednesday	Thursday	Friday	Saturday
1	2	3	4	5	6	7
8	9	10	11	12	13	14
15	16	17	18	19	20	21
22	23	24	25	26	27	28

Review Questions

1. When you look up at the sky, what do you see?

2. Think about the Sun in the sky. Do we see the Sun in the same place all day?

3. The Moon looks different at different times. Tell about the changes. Is there a pattern?

Seasons

In many areas, the **seasons** bring different kinds of weather.

106

Fall

Leaves change color and drop from the trees. Squirrels find seeds to eat. A cool wind blows. We put on our sweatshirts to play in the leaves of fall.

Winter

Trees stand bare. Few animals stir. Snow falls to the ground. We bundle up to keep ourselves warm before we go sledding outdoors.

Spring

Leaves grow on trees. Flowers bloom. Birds are building their nests. The air warms up, and we go out to walk and ride our bikes in the warm, soft breezes of spring.

Summer

The Sun shines brightly on a hot summer day. There isn't a cloud in the sky. The trees give us shade. We can make lemonade. Then, we're off to the beach nearby.

Knowing the weather in each season helps us make choices. We can choose what to wear. We can choose the kind of outdoor activity to do. We can decide to walk or drive to the store.

Review Questions

1. What changes the most from season to season?

2. Each day we choose what to wear. We choose what to do. We choose how to travel. Why is it important to know the weather and the season?

3. What do animals do in different seasons? What did the animals do in the fall, winter, and spring? What do you think they would do in summer?

Getting through Winter

The weather affects many different plants and animals. In spring and summer, the weather is warm. There is plenty of water from rainfall for plants and animals. Plants grow leaves. Some plants grow flowers and fruit, too. Animals can find lots of food to eat.

But in fall, the weather gets cooler. There may be less rainfall. With less water from rainfall, some plants stop growing. Plants need water to grow.

Winter weather can be very cold in some places. Some trees lose their leaves. Many plants become **dormant** with no leaves, flowers, or fruit. Dormant plants are in a resting stage.

When plants are dormant, some animals can't find food. To get through winter, some animals become dormant, too. This is called **hibernation**.

Bears and ground squirrels hibernate. They make dens underground or in a hollow tree. This is where they sleep until spring.

Other animals move for winter. They travel long distances to warmer places where there is food to eat. Animals travel to places where there is more rainfall to provide water to drink. This travel is called **migration**. Migration is a behavior of animals to find food and water.

What animals do you think migrate?

Many kinds of birds migrate. Robins, ducks, and geese migrate. Whooping cranes migrate, too. In spring and summer, the cranes live in Canada. This is where they lay **eggs** and raise their chicks. When it gets cold, the cranes fly south for winter.

Many kinds of **insects** migrate. Dragonflies, ladybugs, and moths migrate. Monarch butterflies migrate, too. In spring and summer, the butterflies live in the northern United States. This is where they lay their eggs. The monarch caterpillars eat the plants that grow there. In fall, many **adult** butterflies fly south, where it is warmer.

To get through winter, some animals hibernate.
Other animals migrate.

What animals in your area migrate?

Review Questions

1. Weather is part of a living thing's environment. Tell about some changes in weather that can affect living things.

2. What do many plants do in the winter?

3. What does *hibernate* mean? What are some animals that hibernate?

4. Do all animals hibernate? If not, what do they do?

Exploring Rocks

Think about a **rock**. A rock has many **properties**. What does the rock look like?

Rocks can be small or large. They can be heavy or light. They can be smooth or rough. They can be round or flat, shiny or dull. Rocks can be different in many ways.

Some rocks are too big to hold in your hand.
A rock can be as big as a mountain!

Other rocks are so small that you can hold thousands in your hand. Look at the picture of a **sand** dune. Can you see the tiny rocks blowing in the wind?

Rocks of all sizes can be found in rivers. Over time, rocks in a river become smooth. Rocks become smooth from rubbing against one another.

Rocks of all sizes can be found in a desert, too.
How big is the rock you're thinking of?

Rocks can be many different colors. They can be black, brown, red, or white. They might even be pink or green. Some rocks have speckles or stripes, too.

Rocks can be many different sizes. They can have different **textures**. They can be many colors and shapes. They can even have patterns.

What does the rock you're thinking about look like?

Review Questions

1. Describe the rock you were thinking about.

2. Where do we find rocks?

3. What are some properties of rocks?

Comparing Solids and Liquids

What is the difference between solids and liquids? They have different properties. Properties describe how something looks or feels.

Shape and size are two properties of solid objects. The shape and size don't change unless you do something to the objects. Solids can be rigid, like a bat. When something is rigid, you can't bend it.

Solids can be flexible, like a sweater. When something is flexible, you can bend and stretch it.

Some solids can be broken into pieces. Each piece has a different shape and takes up less space.

What happens when you put the pieces back together? The solid has the same shape as before. It takes up the same space, too.

Solid objects can be very small, like sand. You can pour sand out of a bucket. But every grain of sand is a solid.

Liquids have properties, too. A liquid can be poured. It doesn't have its own shape. It takes the shape of the container that holds it.

A liquid has a different shape in each different container.

Liquids can be **foamy, bubbly,** or **transparent**. They can be **translucent** or **viscous**.

viscous

translucent

foamy

bubbly and transparent

Solids and liquids are all around you. Can you find the solids in each picture? Can you find the liquids?

Review Questions

1. What are properties?

2. Tell about the properties of solids.

3. Solids can be flexible. What solids in our classroom are flexible?

4. Tell about the properties of liquids.

5. Classify. Is milk a solid or a liquid? Is wood a solid or a liquid?

Heating and Cooling

Have you ever had a glass of lemonade on a hot summer day? After you drink the lemonade, ice is left in the glass. After a while, the ice turns to water. Do you know what happened? The ice **melted**.

When a solid melts, it changes from a solid to a liquid.

Other solids can melt, too. Butter is a solid. But if you **heat** the butter, it melts. Melting changes the physical properties of the butter. Solids melt when they get hot.

Liquids can change to solids, too. Do you know how?

Think about making ice cubes. You pour water in a tray. Then, you put it in a cold place. When the water gets very cold, you have solid ice cubes! When a liquid **cools** or **freezes**, it changes to a solid.

Can you think of other liquids that turn to solid?

Liquid chocolate turns to solid as it cools. Liquid chocolate can be poured into molds. When the chocolate cools, it is solid. That's how candy is made into different shapes!

This candle is melting and freezing at the same time. The wax near the flame gets hot. It melts and turns to liquid. The liquid wax flows down the side of the candle. When the liquid wax is away from the heat, it cools and freezes.

Review Questions

1. Tell about how solids change into liquids.

2. Tell about how liquids change into solids.

3. The ice in the picture is changing. It is melting. Why?

4. Melting changes the physical properties of ice. How could you change paper? How could you change wood?

Resources

Things that people use are resources. Resources from Earth are natural resources. Resources that people change to make other things are **human-made resources**.

Plants and animals are natural resources. Wood from trees is a natural resource. People change wood from trees into lumber, cardboard, and paper. These things are human-made resources.

Cotton plants are a natural resource. People use cotton to make thread. The thread is used to make cotton fabric for clothes. Cotton clothes are good for warm and sunny days.

Wheat plants are a natural resource. Wheat seeds are harvested. People use wheat to make cereal and flour for bread and other foods.

Sheep are a natural resource. People use the fur from sheep to make wool thread. The thread is used to make wool fabric for clothes. Wool clothes are good for cold and snowy days.

Water is a natural resource. People use fresh water from rivers and lakes. The water is treated to make it clean for people to use. It is used for drinking, washing, and cooking.

Earth materials like gravel, sand, and clay are natural resources. People use gravel and sand to make concrete and asphalt. People use clay to make bricks and pottery.

Air is a natural resource. Plants and animals need air to live.

The Sun is a natural resource. The Sun heats the air, land, and water. The Sun makes our weather, too.

Look at the pictures. Which things are natural resources? Which things are resources made by people?

So many of the materials we use come from nature. People must keep these natural resources safe. We must use them wisely. We must **conserve** them.

Conservation can help keep air, water, and soil clean. Conservation can make resources last longer. People can conserve resources if they do the three Rs: reduce, reuse, and recycle.

Reduce means to use less of something. People can use less water. Companies can use less packaging for the products they make.

Reuse means to use again. People can reuse containers instead of throwing them away. Broken objects can be fixed instead of replaced. We can give outgrown clothes to others.

Recycle means to turn old things into material for new things. For example, the plastic in many bottles can be recycled. It can be made into new bottles, carpeting, or cloth. We can also recycle paper, metal, and glass.

Review Questions

1. Trees are a natural resource. What human-made resources come from trees?

2. Gravel and sand are natural resources. What human-made resources come from them?

3. Look at pages 156–157. What are the natural resources? What are the human-made resources?

4. How can we conserve natural resources?

Life Science
Insects and Plants

Table of Contents

Animals and Plants in Their Habitats **167**
Flowers and Seeds **183**
So Many Kinds, So Many Places **192**
Variation **198**
Insect Shapes and Colors **205**
Insect Life Cycles **212**
Life Goes Around **221**
Fossils **235**

Animals and Plants in Their Habitats

Look at this grassland **habitat**. Do you see anything **living** here? Let's take a closer look.

Look! A grasshopper sits in the green grass. Grass plants and grasshoppers are living in this grassland habitat. Grass and grasshoppers get their **basic needs** met in the grassland habitat.

Land animals need food, water, **air**, space, shelter, and a comfortable **temperature**. They depend on their **environment** for the things they need. Grasshoppers eat grass for food and water. In the grass, grasshoppers have air and shelter. In summer, the grassland provides a warm habitat for **insects**, like grasshoppers, and other animals, like prairie dogs.

Plants need water, air, sunlight, and **nutrients**. Plants depend on their environment for these things. Rainfall provides water to the soil. Grass gets water and nutrients from the rich, moist grassland soil. Their roots take up water and nutrients. Without rainfall, the grass would not grow. Air flows around the grass. Grass uses sunlight to make the food it needs to live.

Other parts of the country are different from the grassland habitat. This is a desert habitat. What do you see living here? Let's take a closer look.

Plants and animals are living in the desert, too. How do they get the things they need to live?

Desert plants get water and nutrients from the desert soil. The plants have large root systems to collect water. Some desert plants, like cactuses, have thick spongy stems. Cactus stems store water for the plant to use later.

The **Sun** shines brightly on the desert most of the time. Air flows easily around the desert plants. Deserts are very hot during the summer. But desert plants can live even in very hot temperatures.

Some desert animals eat other animals for food. Some desert animals eat **seeds** for food. Harvester ants gather seeds. They store the seeds to eat during the year. Ants get water from the seeds they eat.

Ants dig tunnels and chambers. Their underground tunnels give ants a safe space to live. The tunnels provide shelter for the ants so they can **survive** the hot desert summer.

This is a tundra habitat in winter. Some parts of Alaska have this kind of habitat. What do you see living here? Let's take a closer look.

Tundra plants can survive cold temperatures. Tundra plants survive the winter in a resting stage. When the Sun warms the land, the plants begin to grow again. The plants take up water and nutrients from **melting** ice and **snow**. They use sunlight to make their food.

Caribou roam the open space of the tundra. They eat the short plants. They drink from pools of melted snow. Caribou have thick fur to protect them from the cold temperature. But the fur is not thick enough to protect them from mosquitoes.

Mosquitoes burrow down in tundra plants in the fall. They rest there during the cold winter. When the snow melts in spring, the mosquitoes look for food and mates. The female mosquitoes lay **eggs** on pools of water.

After the eggs hatch, mosquito **larvae** eat tiny bits of food in the water. The mosquito larvae grow quickly for a few weeks. Then they swim to the surface of the water and break out of their old skin. Now they are **adult** mosquitoes and fly to find food.

When you look closely at a habitat, you will find animals and plants living there. Living things **thrive** when they have what they need to live.

Review Questions

1. What do animals need to live?

2. What do plants need to live?

3. Living things in the tundra depend on each other. They also depend on things like sunlight, water, and soil. Give examples. Compare.
   a. What does a caribou depend on?
   b. What does a tundra plant depend on?
   c. How are caribou and tundra plants alike? How are they different?

4. Some living things in the tundra go into a resting stage. How does this help them?

Flowers and Seeds

These are wild brassica plants. Each plant grows a lot of **flowers**. But brassica plants do not grow flowers to look pretty. The flowers are an important part of the plant's **life cycle**.

Soon, the flowers fade and dry up. Something new appears right where each flower once grew. It looks like a little green bean. It is a seedpod.

Weeks later, the seedpods are big and dry. There are about six seeds inside each seedpod. What do you think will happen if someone plants the new seeds?

Brassica plants are not the only plants that make seeds. Cherry trees make seeds. Where are they found?

There is one seed inside each cherry. And where does the cherry grow? It grows right where the cherry flower was.

Plants grow flowers. The flowers grow into **fruit**. Fruit have seeds inside. When the seeds grow into new plants, it is called **reproduction**.

Apple tree

Apple flowers

Apple fruit

Apple seeds

Have you ever seen tomato flowers? Tomato flowers grow into tomatoes. Tomatoes are fruit. They have seeds.

Can you see the strawberry flowers?
Strawberry flowers grow into fruit.
Strawberries have seeds, too.

New plants grow from seeds. Seeds are found in fruit. Fruit grow out of flowers. Flowers and fruit are important in the life cycles of plants.

Review Questions

1. Name one plant. Tell about its flowers.

2. Where are the seeds on a full-grown brassica plant?

3. Name two fruits you like to eat.

4. Four parts of a plant are important for its life cycle. Name them.

So Many Kinds, So Many Places

This amazing animal is an insect. Flies, ants, and crickets are all insects, too. There are so many kinds of insects. Insects are everywhere! Can you name some others?

No matter where you are, an insect is probably near you. Insects are in the air and in the water. Some creep in the Arctic snow. Others scamper around in the desert.

These ladybugs have gathered on a tree trunk. Some insects live on the tops of mountains. Other insects live in the rain forest. Insects are everywhere!

Insects might seem like pests to you. Some insects eat clothes, buildings, and crops. But insects are very important. Many different animals depend on them for food.

Insects are important for plants and people, too. Bees visit flowers, and that helps fruit grow. They also make sweet honey.

People use thread from the cocoon of the silkworm to make clothing.

Next time you go outside, look for insects. They are everywhere!

Review Questions

1. Living things depend on each other. Give examples. Compare.
    a. How are insects useful to animals?
    b. How are insects useful to plants?

2. How can insects cause problems?

Variation

Variation means difference. When we looked at darkling beetles, we saw variation. They were not all the same. Do other **organisms** have variation?

People are different. Some people are short. Some have brown eyes. Some have black hair. Some have freckles. Everyone is different when you look closely enough.

These black bears have the same shape and size.
But like other animals, they have variation too.
What kind of variation do you see?

Trout have color variation, too. Some of them are silver. Others are brightly colored. Some have a lot of spots. Others have only a few spots. Trout have both color and pattern variation.

These shells are all from the same kind of scallop. What kinds of variation can you see?

Here is a garden with marigolds. Do you see any variation?

Review Questions

1. Name five variations you can observe about people.

2. Tell about variations in trout.

3. Think of another animal. Tell about its variations.

Insect Shapes and Colors

Insects are different shapes and colors. The shape or color can help insects hide. An enemy might not see an insect that looks like its habitat. A hungry bird or lizard might think this insect is a leaf.

This praying mantis hides in the leaves, waiting to catch an insect to eat.

The walking sticks on this twig are very hard to see. Can you find them?

Look at the bright colors and design on this ladybug beetle. Do you think it is hiding from its enemies?

Some insects are very easy to see. They are very colorful. They might have special markings.

Brightly colored insects often taste bad. They make other animals sick. Animals learn to stay away!

The spots on this beetle look like huge eyes. A hungry animal might think the beetle is a much bigger insect. The animal might be scared away.

Review Questions

1. Why are some insects hard to see?

2. Why are some insects so colorful?

3. How can body shape or color help an insect stay alive?

Insect Life Cycles

Insects might look different at each stage of their lives. Most insects go through four stages. The stages are egg, larva, **pupa**, and adult. The eggs of this insect were laid inside cells.

After a few days, a larva hatches from each egg. The tiny larva stays curled up inside the cell. The larva eats food made from pollen and honey. This food makes the larva grow.

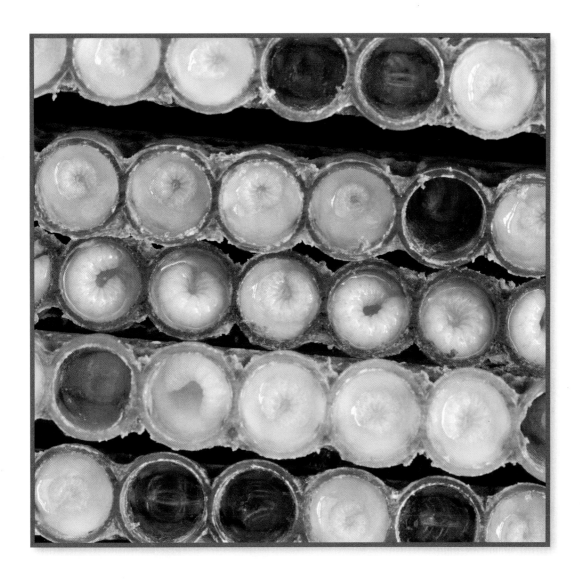

Then, the cell is covered with wax. Inside the cell, each larva turns into a pupa.

In the pupa stage, the insect goes through a big change. Soon, an adult crawls out of each cell. Do you know what insect this is?

It's a bee! After a short rest, the bee can go right to work. Young adult bees work in the hive. Older bees work outside the hive.

The larvae of different insects do not look the same. These larvae will become insects you know well. What will they look like as adults?

Moths and mosquitoes!

Some kinds of insects don't have larvae or pupae. When they hatch from eggs, they are called **nymphs**. Many nymphs look like their parents, but smaller.

Milkweed bugs go through four nymph stages. In each new stage, they look more like an adult. How many different nymph stages can you find?

Review Questions

1. Tell about the life cycle of a bee.

2. Tell about the life cycle of a milkweed bug.

Life Goes Around

On a lucky day, you might see a ladybug. A ladybug is red with black spots. This is an adult ladybug. But have you ever seen a baby ladybug?

Adult ladybugs mate. Then, the female lays eggs. When an egg hatches, a larva comes out. The black larva is a baby ladybug. But it doesn't look like its parents. The larva eats and grows for about 4 weeks.

Then, the larva **pupates**. Inside the pupa, the larva is changing. When the pupa opens, an adult ladybug comes out. It is red with black spots. Now it looks just like its parents.

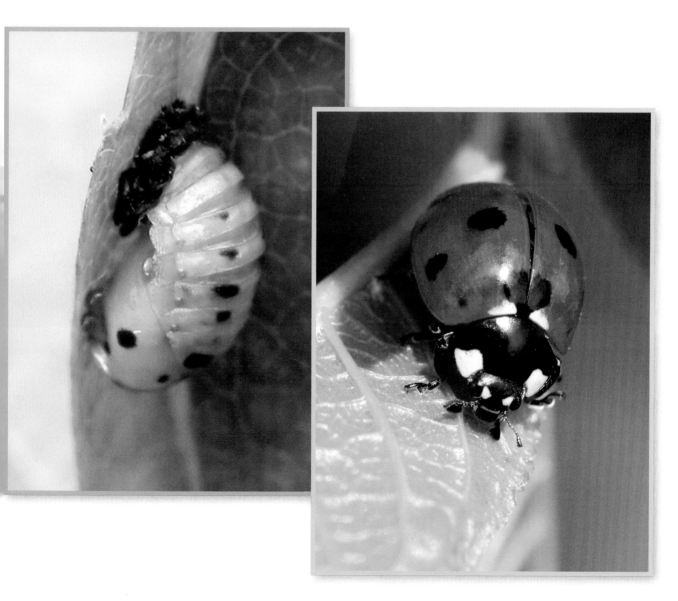

The ladybug life cycle is like the life cycle of many other insects. It is like the life cycle of mealworms. It is like the life cycle of butterflies and moths. But it is different from the life cycle of some other animals.

Some animals hatch from eggs. Some animals are born alive. They all grow up to be adults. The adults mate and have babies called **offspring**.

Every animal goes around the life cycle. Cycle means to go around. The life cycle of a robin looks like this.

Robin eggs

Baby robins

Young robin

Adult robin

Trout lay eggs in streams. After 6 to 8 weeks, the eggs hatch. Tiny, fat babies swim out. You can see that they are fish. But they don't look like their parents yet.

For the next year, they grow up little by little. In 2 years, they are adults. They look just like their parents. They mate and lay eggs in streams. Can you describe the trout life cycle?

Frogs lay eggs in water, too. When an egg hatches, a tadpole swims out. It looks more like a fish with a big head than a frog. It doesn't look like its parents yet.

The tadpole eats and grows. In a few weeks, the tadpole starts to change. Its long, flat tail gets shorter. Its legs start to grow.

In a few more weeks, the tadpole has grown into a frog. It looks just like its parents. Can you describe the frog life cycle?

Ducks lay eggs in a nest in a **marsh**. The mother duck sits on the eggs to keep them warm. When they hatch, the babies are fluffy and yellow. The babies are called ducklings. You can see that they are ducks. But they don't look like their parents yet.

The ducklings eat and grow. In a few weeks, they get their brown feathers. In a few months, they are adults. They look just like their parents. In the next year, the adult ducks will mate. They will raise new families of ducklings. Can you describe the duck life cycle?

Mice do not lay eggs. Baby mice grow inside the mother. The babies are born alive. Newborn mice are pink, hairless, and blind. You can see that they are mice. But they don't look like their parents yet.

In a few days, the babies open their eyes. Their fur starts to grow. In a few months, they will be adults. They will be ready to continue the life cycle. They will have babies of their own. Can you describe the life cycle of mice?

Review Questions

1. Does a ladybug larva look like its parents?

2. Tell about the life cycle of a ladybug.

3. Tell about the life cycle of a different animal.

4. Name five animals that hatch from eggs.

5. Name three animals that are born alive (not from eggs).

Fossils

How do we know what plants and animals looked like millions of years ago? We look at **fossils**. Fossils are the remains of plants and animals that lived a long time ago. Scientists study fossils to learn about the past.

Dinosaurs lived a long, long time ago. No dinosaurs are living today. But scientists can study dinosaur fossils to learn about them.

One of the most famous dinosaur fossils is named Sue. Sue is a *Tyrannosaurus rex*. This is what Sue might have looked like when she was found in South Dakota.

Scientists dug Sue out of the ground very carefully.

In the lab, Sue's fossil bones were carefully cleaned.

After a lot of hard work, all of Sue's bones and teeth were ready to put together.

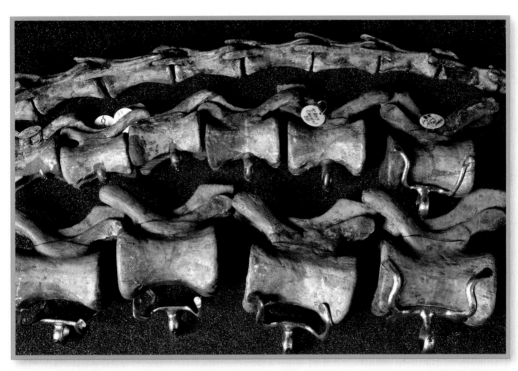

It took a long time to get all the fossil bones in the right places. Finally, the bones all fit together. Everyone can now see what Sue's skeleton looked like.

After the skeleton was together, scientists wanted to know what Sue looked like when she was alive. They used pretend muscles, skin, and eyes on a copy of her skeleton to make her look real.

This is what Sue might have looked like.

Dinosaurs have not lived on Earth for millions of years. But animals that look like dinosaurs are found on Earth today. Lizards, such as iguanas, look like dinosaurs.

Insects also lived when dinosaurs roamed Earth. Fossils of insects are found in **rock**. Insect fossils are found in **amber**, too.

Amber is very old **resin** from a tree. Resin is a very sticky **liquid** that comes from some kinds of plants. If you have ever touched pine tree sap, then you know how sticky resin can be.

If an insect landed in resin, it was trapped and died. Millions of years later, those insects are fossils in amber. Bees, dragonflies, and wasps are some of the insect fossils found in amber.

Review Questions

1. What are fossils?

2. Where are fossils found?

3. What do fossils tell us?

4. Why are fossils important?

References

Table of Contents

Science Safety Rules **248**
Outdoor Safety Rules **249**
Tools for Scientific Investigation. **250**
Glossary. **266**
Photo Credits. **272**

Science Safety Rules

1. Listen carefully to your teacher's instructions. Follow all directions. Ask questions if you don't know what to do.
2. Tell your teacher if you have any allergies.
3. Never put any materials in your mouth. Do not taste anything unless your teacher tells you to do so.
4. Never smell any unknown material. If your teacher tells you to smell something, wave your hand over the material to bring the smell toward your nose.
5. Do not touch your face, mouth, ears, eyes, or nose while working with chemicals, plants, or animals.
6. Always protect your eyes. Wear safety goggles when necessary. Tell your teacher if you wear contact lenses.
7. Always wash your hands with soap and warm water after handling chemicals, plants, or animals.
8. Never mix any chemicals unless your teacher tells you to do so.
9. Report all spills, accidents, and injuries to your teacher.
10. Treat animals with respect, caution, and consideration.
11. Clean up your work space after each investigation.
12. Act responsibly during all science activities.

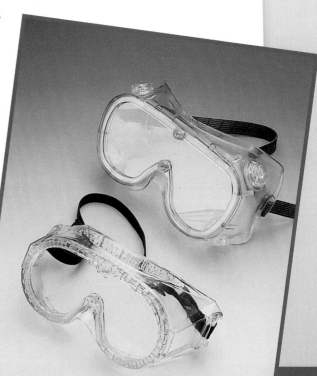

Outdoor Safety Rules

1. Listen carefully to your teacher's instructions. Follow all directions. Ask questions if you don't know what to do.

2. Never put any materials in your mouth.

3. Tell your teacher if you have any allergies. Let your teacher know if you have never been stung by a bee.

4. Dress appropriately for the weather and for the outdoor experience. For long activities in the Sun, wear long sleeves, long pants, and a hat. Use sunscreen.

5. Stay within the designated study area and with your partner or group. When you hear the "freeze" signal, stop and listen to your teacher.

6. Never look directly at the Sun or at the sunlight being reflected off a shiny object.

7. Most plants and animals in the schoolyard are harmless. Know what the skin-irritating plants in your schoolyard look like, and do not touch them. Ask your teacher if you don't know.

8. When looking under a stone or log, lift the side away from you so that any living things can move away from you.

9. If a stinging insect is near you, stay calm and slowly walk away from it. Tell your teacher if you are stung or bitten by any living thing.

10. Take good care of the outdoor environment, and respect all living things. Never release any living things into the environment unless you collected them there.

11. Always wash your hands with soap and warm water after handling plants, animals, and soil.

12. Return to the classroom with all of the materials you brought outside.

Tools for Scientific Investigation

During science class, you use a lot of different tools. These tools help you collect information. They help you record and compare information, too. You use the same tools that scientists do!

Notebooks

This is a notebook. You can use a notebook to record observations, data, and explanations.

Computers

This is a computer. You can use a computer to collect, record, and organize data. A computer can help you compare information, too.

Safety Goggles

This is a pair of safety goggles. You wear safety goggles to protect your eyes while doing science.

Measurement Tools and Rulers

You can use objects as units to compare the length of objects. You can measure the length of a piece of wood by comparing it to the length of a paper clip, a clothespin, or string.

You can use a ruler to measure and compare the length of objects, too.

Balances

This is a balance. You can use a balance to compare the weight of two objects. Using standard units, you can measure the mass of a single object.

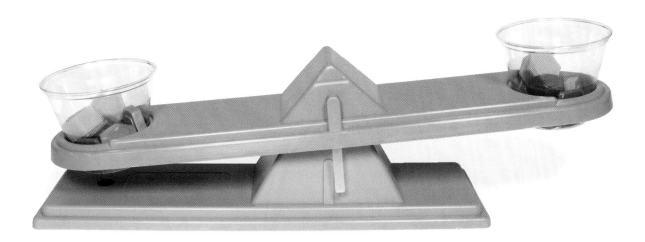

Timing Devices

These are two kinds of devices to keep track of time. You can use a clock or a stopwatch to measure how long it takes for something to happen.

Beakers

This is a beaker. You can use a beaker to measure liquids.

Cups

This is a cup. You can use a cup to collect things, such as rocks and soil samples. You can use a cup to observe things, to mix things, and even to grow plants.

Bowls

This is a container, or bowl. You can use a bowl to collect things, such as water and pebbles. You can use a bowl to mix things, too.

Weather Instruments

This is a rain gauge.
A rain gauge measures rain or snow.

This is a thermometer.
A thermometer measures temperature.

This is an anemometer. An anemometer measures wind speed.

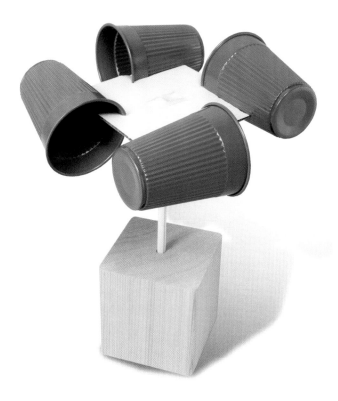

Hand Lenses

This is a hand lens. You can use a hand lens to observe living things (plants and animals) and objects up close.

Magnets

These are magnets. You can use a magnet to explore the properties of materials and objects.

Collecting Nets

This is a collecting net. You can use a collecting net to collect small animals from a pond or aquarium.

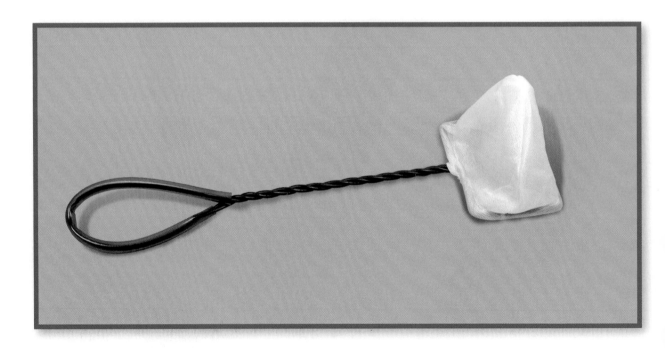

Habitats

This is an aquarium. An aquarium is a small water habitat. You can observe animals and plants in an aquarium.

This is a terrarium. A terrarium is a small land habitat. You can observe plants and animals in a terrarium.

Glossary

adult a fully grown organism (119, 180)

air a mixture of gases that we breathe (20, 47, 169)

amber a hard fossil resin from a tree (243)

axis a straight line around which something turns (24)

balance to be in a stable position (9)

basic need something that is needed for plants and animals to survive. Plants and animals need air, water, food, space, shelter, and light. (168)

bubbly describes a liquid that is full of bubbles (136)

cloud a group of very small water drops in the sky. Cirrus, cumulus, and stratus are kinds of clouds. (58)

condensation when a gas changes to a liquid (73)

conserve to keep resources safe and use them wisely (158)

cool to make something colder (143)

counterbalance to place weights on an object to keep it in a stable position (15)

dormant sleeping or not growing (115)

egg the first stage of a life cycle (118, 179)

energy the ability to make things move or change. Where there is motion, sound, heat, or light, there is energy. (41)

environment all the things and conditions around a living thing. Soil, air, water, and other plants or animals can be part of an environment. (169)

evaporation when a liquid changes to a gas **(72)**

flower the part of a plant that grows into fruit **(183)**

foamy describes a liquid that has a layer of bubbles on top **(136)**

force a push or a pull **(19)**

fossil a part of a plant or animal that lived long ago. Fossils can be bones, shells, or leaves. Fossils can also be tracks or burrows of past life on Earth. **(235)**

freeze to change a liquid to a solid by cooling it **(143)**

fresh water water without salt. Fresh water is found in streams, lakes, and rivers. **(78)**

fruit the part of a plant with seeds in it. Flowers grow into fruit, and fruit produce seeds in plant reproduction. **(187)**

gas matter that can't be seen but is all around. Air is an example of a gas. **(48)**

gravity a force that pulls things toward Earth **(21)**

habitat the place or natural area where plants or animals live **(167)**

heat to make something warmer **(142)**

hibernation when animals sleep through the winter **(116)**

human-made resource a material made by people that we need or use. Paper, cloth, and bricks are human-made resources. **(147)**

hurricane a strong, wet, and windy storm that forms over warm ocean water **(90)**

insect an animal that has six legs and three main body parts. They are the head, thorax, and abdomen. **(119, 169)**

larva (plural larvae) a stage in the insect life cycle after hatching from eggs. Insect larvae look different from their parents and are often wormlike. **(180)**

life cycle the stages in the life of a plant or animal **(183)**

liquid matter that flows freely and takes the shape of its container **(48, 244)**

living alive **(167)**

marsh soft, wet land that is sometimes covered with water **(82, 230)**

matter anything that takes up space **(47)**

measure to find the amount of something **(87)**

melt to change a solid to a liquid by heating it **(141, 177)**

meteorologist a person who studies the weather **(87)**

migration when animals move when the season changes **(117)**

Moon the object we see in the night sky and sometimes during the day. Some of the Moon shapes we observe and describe are the full Moon, crescent Moon, quarter Moon, and gibbous Moon. **(100)**

motion the act of moving. Motion happens when something changes position **(9)**

natural resource a supply of something in nature. The Sun, soil, and air are natural resources. **(80)**

nutrient something that living things need to grow and stay healthy **(170)**

nymph a stage in the insect life cycle that has no larva or pupa. Nymphs look like their parents, but are smaller. **(218)**

offspring a new plant or animal produced by a parent **(225)**

organism a living thing. Plants and animals are organisms. **(198)**

pitch how high or low a sound is **(42)**

position where something is **(34)**

precipitation rain, snow, or hail falling from the clouds **(75)**

property something you can observe about an object or a material. Size, color, shape, texture, and smell are properties. Volume is an example of a property of sound. **(41, 122)**

pull when you make things move toward you. Pulling is a force. **(19)**

pupa (plural pupae) a stage in the insect life cycle between the larva and adult stages **(212)**

pupate to change into a pupa **(223)**

push when you make things move away from you. Pushing is a force. **(9, 53)**

rain one kind of weather that falls from the clouds as water drops **(58)**

reproduction the process of producing offspring **(187)**

resin a sticky liquid that comes from some plants **(244)**

rock a solid earth material. Rocks are made of minerals. **(122, 243)**

roll to move from one place to another by turning over and over **(29)**

salt water water with salt. Salt water is found in seas and the ocean. **(81)**

sand rocks that are smaller than gravel, but bigger than silt **(125)**

season one of four times of year that has different weather. Winter, spring, summer, and fall are seasons. **(106)**

seed the part of a plant found inside fruit. Seeds can grow into new plants. **(174)**

shadow a dark area made by blocking the light from the Sun **(99)**

sound something you hear **(38)**

snow one kind of weather that happens when it is very cold. Frozen water falls from clouds. **(63, 177)**

solid matter that holds its own shape and always takes up the same amount of space **(48)**

spin to move by turning around an axis **(9)**

stable position steady, not falling over **(15)**

state one of the three groups of matter: solid, liquid, or gas **(48)**

star an object in the sky that makes light and heat **(93)**

storm weather that has strong winds and can bring rain or snow **(67)**

Sun a star we see in the day sky. The Sun warms the land, air, and water. **(61, 173)**

survive to stay alive **(175)**

temperature a description of how hot or cold something is **(60, 169)**

texture the way something feels **(129)**

thrive to grow fast and stay healthy **(181)**

tornado a twirling, cloudy, dangerous storm **(89)**

translucent describes a liquid or solid that is clear enough to let light through but is not clear enough to see something on the other side **(136)**

transparent describes a liquid or solid that you can see through easily **(136)**

variation difference **(198)**

vibration a back-and-forth motion. Vibration makes sound. **(39)**

viscous describes a liquid that is thick and slow moving **(136)**

volume how soft or loud a sound is **(41)**

water cycle when water from Earth evaporates, condenses into clouds, and falls back to Earth as precipitation **(75)**

water vapor water as a gas **(72)**

weather the condition of the air outdoors **(58)**

weather balloon a balloon that carries weather instruments into the sky **(88)**

wind moving air **(52)**

Photo Credits

Cover/1: Lauren Blackwell/Shutterstock; 2/24: iStockphoto/Frans Rombout; 3/147: iStockphoto/GaryAlvis; 4/207: James Laurie/Shutterstock; 5: Ivancovlad/Shutterstock; 7: Ruta Saulyte-Laurinaviciene/Shutterstock; 8: iStockphoto/Bryan Regan; 9: Alloy Photography/Veer; 10: Jack schiffer/Shutterstock (l); PavleMarjanovic/Shutterstock (r); 11: Renata Osinska/Shutterstock; 12: iStockphoto/Jiang Dao Hua (l); iStockphoto/Stephanie Kennedy (r); 13: charles taylor/Shutterstock; 14: Pfong001/Dreamstime.com; 15: John Quick; 16: ID1974/Shutterstock (l); design56/Shutterstock (tr); Great Divide Photography/Dreamstime.com (m); Mike Price/Shutterstock (br); 17: JEO/Shutterstock (tl); Kanwarjit Singh Boparai/Shutterstock (bl); iStockphoto/Alexey Sokolov (tr); iStockphoto/paul kline (br); 18: iStockphoto/alain cassiede; 19: iStockphoto/visual7; 20: Marmi/Shutterstock; 21: Ariel Skelley/CORBIS; 22: iStockphoto/Nicolas McComber; 23: Cristy/Shutterstock; 25: Goldution/Shutterstock; 26: Olaru Radian-Alexandru/Shutterstock (l); BEPictured/Shutterstock (tr); Africa Studio/Shutterstock (m); iStockphoto/Andy Dean (br); 27: Robert Hackett/Shutterstock (tl); Le Do/Shutterstock (bl); Katrina Leigh/Shutterstock (tr); dragon_fang/Shutterstock (br); 28: Don Tran/Shutterstock; 29: iStockphoto/Alf Ertsland; motorolka/Shutterstock (i); 30: Jan Kruger/Getty Images; John Quick (l); 31–33: John Quick; 34–35: Laurie Meyer; 36: John Quick (l); Margo Sokolovskaya/Shutterstock (r); 37: nikkytok/Shutterstock; 38: iStockphoto/John Stelzer; 39: Zepherwind/Dreamstime.com; 40: Pedro Talens Masip/Shutterstock; 41: iStockphoto/naphtalina; 42: Mike Kemp/Getty Images; 43: iStockphoto/Minnie Menon (l); eAlisa/Shutterstock (m); John McLaird/Shutterstock (r); 45: hfng/Shutterstock; 46: Thomas Barrat/Shutterstock; 47: Mary Terriberry/Shutterstock; 48: GSPhotography/Shutterstock; 49: nmedia/Shutterstock; 50: Ricardo Garza/Shutterstock; 51: Orientaly/Shutterstock; 52: BestPhotoByMonikaGniot/Shutterstock; 53: Ken Seet/Corbis (b); 53: Ljansempoi/Dreamstime.com; 54: iStockphoto/Michael Hoefner; 55: iStockphoto/Bevs Photos; 56: gary718/Shutterstock; 57: Brandon Seidel/Shutterstock; 59: iStockphoto/Alberto Pomares (t); iStockphoto/Blueberries Advertising (m); Ralph Loesche/Shutterstock (b); 60: Yuriy Kulyk/Shutterstock; 61: iStockphoto/Craig Veltri; 62: Dudarev Mikhail/Shutterstock; 63: Gordon Warlow/Shutterstock; 64: iStockphoto/Daniel Halvorson; 65: iStockphoto/Eric Gevaert; 66: Gary Blakeley/Shutterstock; 67: Alexander Studentschnig/Shutterstock; 68: Tony Freeman/PhotoEdit; 69: Rudy Lopez Photography/Shutterstock; 70: iStockphoto/Sven Klaschik; Yuriy Kulyk/Shutterstock (i); 71: Triff/Shutterstock; 72: iStockphoto/David Sucsy; 73: Eising FoodPhotography/StockFood; 74: Antonio V. Oquias/Shutterstock; 75: elwynn/Shutterstock; 76: Aliaksei Lasevich/Shutterstock; 77: Tatiana Grozetskaya/Shutterstock; Blend Stock Photos/Fotosearch (i); 78: iStockphoto/Kent Metschan; Lawrence Hall of Science (i); 79: iStockphoto/ray roper (t); Lawrence Hall of Science (b); 80: Ppaauullee/Shutterstock (tl); spe/Shutterstock (tr); iStockphoto/Michelle Gibson (bl); iStockphoto/Rob Friedman (br); 81: Iakov Kalinin/Shutterstock; 82: Sarah Pettegree/Shutterstock; 83: Ricardo Garza/Shutterstock; 84: iStockphoto/Alberto L. Pomares; 85: Lawrence Hall of Science; 86: Lawrence Hall of Science (tl/br); Ricardo Garza/Shutterstock (tr); Bonita R. Cheshier/Shutterstock (bl); 87: Stephen J. Krasemann/Photo Researchers, Inc.; 88: Mark C. Burnett/Photo Researchers, Inc.; 89: iStockphoto/victor zastol`skiy; 90: Mayra Pau/Photos.com; 91: iStockphoto/Chan Pak Kei; 92: mangojuicy/Shutterstock; 93: Yuriy Kulyk/Shutterstock; 94: Boris Franz/Shutterstock; 95: Albert Cheng/Shutterstock (t); Vibrant Image Studio/Shutterstock (b); 96: Cynthia Kidwell/Shutterstock; 97: Igordabari/Dreamstime.com (t); oriontrail/Shutterstock (b); 98: James Peragine/Shutterstock; 99: kwest/Shutterstock; 100: ifong/Shutterstock; 101: MountainHardcore/Shutterstock; 102: Terrance Emerson/Shutterstock; 103: iStockphoto/Shaun Lowe; 104: Lawrence Hall of Science; 105: Steven Bostock/Shutterstock; 106: Ales Nowak/Shutterstock, Sandra Cunningham/Shutterstock (t); sonya etchison/Shutterstock (b); 107: Richard Hutchings/PhotoEdit; Bruce MacQueen/Shutterstock (i); 108: Barbara Stitzer/PhotoEdit; Heidi Brand/Shutterstock (i); 109: Dean Mitchell/Shutterstock; iStockphoto/Jill Battaglia; 110: Bill Bachmann/PhotoEdit; iStockphoto/Morgan Lane Studios (i); 111: Studio 1One/Shutterstock (b); nolie/Shutterstock (m); aarrows/Shutterstock (b); 112–113: Jemini Joseph/wildbirdimages.com; 114: iStockphoto/TT; JinYoung Lee/Shutterstock (i); 115: iStockphoto/Andrey Stepanov; Pawel Kielpinski/Shutterstock (i); 116: Stephen Lang/Visuals Unlimited, Inc.; 117: Delmas Lehman/Shutterstock; 118: iStockphoto/jim kruger; iStockphoto/Robert Blanchard (i); 119: iStockphoto/Simon Phipps; Robert Crow/Shutterstock (i); 120: Chasbrutlag/Dreamstime.com (tl); Joy Brown/Shutterstock (tr); iStockphoto/Missing35mm (bl); Gerald A. DeBoer/Shutterstock (br); 121: BMJ/Shutterstock; 122: ZouZou/Shutterstock; 123: Merydolla/Shutterstock; 124: Jim Feliciano/Shutterstock; 125: iStockphoto/Kurt Cotoaga; Leo Kenney (i); 126: wong yu liang/Shutterstock; 127: Paul B. Moore/Shutterstock; 128: RonGreer.Com/Shutterstock; 129: Lawrence Hall of Science; 130: Roni Lias/Shutterstock; 131: Jeff Thrower/Shutterstock; 132: iStockphoto/Nicole S. Young; 133: Serdar Tibet/Shutterstock; 134: Shannon Fagan/Getty Images; 135: John Quick; 136: John Quick (t); iStockphoto/draschwartz (b); 137: iStockphoto/Khanh Trang (tl); iStockphoto/Joao Virissimo (tr); iStockphoto/Ljupco Smokovski (bl); 138: Thomas M. Perkins/Shutterstock; Nayashkova Olga/Shutterstock (b); 139: Elke Dennis/Shutterstock (t); Monkey Business Images/Shutterstock (b); 140: iStockphoto/MBPhoto; 141: Dave Bradley Photography, Inc.; 142: Joy Brown/Shutterstock; Witold Ryka/Shutterstock (i); 143: Dave Bradley Photography, Inc.; 144: iStockphoto/Gustavo Andrade (l); Baloncici/Shutterstock (r); 145: audioscience/Shutterstock; 146: iStockphoto/Erikki Makkonen; 148: LianeM/Shutterstock; iStockphoto/Olivier Blondeau (i); 149: GONUL KOKAL/Shutterstock; Studio 1One/Shutterstock (i); 150: Eric Gevaert/Shutterstock; Monkey Business Images/Shutterstock (i); 151: Claudia Steininger/Shutterstock; Digital Vision/Getty Images (i); 152: iStockphoto/Keith Webber Jr.; Péter Gudella/Shutterstock (i); 153: Dianne Maire/Shutterstock; Jose Gil/Shutterstock (i); 154: bcampbell65/Shutterstock; 155: Emanuel/Shutterstock; 156: Dmitry Kalinovsky/Shutterstock (tl); iStockphoto/DNY59 (tr); Laurent Renault/Shutterstock (m); bierchen/Shutterstock (bl); Robert Milek/Shutterstock (br); 157: Alexandr Makarov/Shutterstock (tl); iStockphoto/Dagmara Ponikiewska (tr); Mikhail Levit/Shutterstock (m); ilker canikligil/Shutterstock (bl); Christopher Jones/Shutterstock (br); 158: Miao Liao/Shutterstock; 159: Africa Studio/Shutterstock, Rtimages/Shutterstock (i); 160: B Calkins/Shutterstock; 161: mangostock/Shutterstock; 162: Dgrilla/Shutterstock (l); Mariyana Misaleva/Shutterstock (tr); Venus Angel/Shutterstock (br); 163: Morgan Lane Photography/Shutterstock; 165: iStockphoto/Eva Kaufman; 166: Rob Hainer/Shutterstock; 167: WEN BILLY/Shutterstock; 168: iStockphoto/André Gonçalves; 169: Peter Kirillov/Shutterstock; 170: Maksym Protsenko/Shutterstock, NatUlrich/Shutterstock (i); 171: Jayne Chapman/Shutterstock; 172: EuToch/Shutterstock; 173: Anton Foltin/Shutterstock; 174–175: Alex Wild/alexanderwild.com; 176: Dominik Michalek/Shutterstock; 177: Larry Malone/Lawrence Hall of Science; 178: iStockphoto/Paul Wilson; 179: iStockphoto/Jan Rihak; 180: iStockphoto/Douglas Allen (l); Henrik Larsson/Shutterstock (r); 181: Stephanie Coffman/Shutterstock; 182: Jeff Banke/Shutterstock; 183: Karin Hildebrand Lau/Shutterstock; iStockphoto/Cecelia Sullivan (i); 184: Tierbild/OKAPIA/Photo Researchers, Inc.; 185: Thomas W. Martin/Photo Researchers, Inc.; 186: Lakhesis/Shutterstock (l); David Orcea/Shutterstock (r); 187: LOYISH/Shutterstock (tl); BrankaVV/Shutterstock (tr); Mazzzur/Shutterstock (br); Rob Stark/Shutterstock (bl); 188: Denis and Yulia Pogostins/Shutterstock (t); Peter zijlstra/Shutterstock (b); 189: LilKar/Shutterstock (b); Picsfive/Shutterstock (b); 190: Mircea BEZERGHEANU/Shutterstock; 191: Ben Zastovnik/Shutterstock (t); photomaru/Shutterstock (bl); Korolevskaya Nataliya/Shutterstock (r); 192: Miroslav Hlavko/Shutterstock; 193: vblinov/Shutterstock; iStockphoto/Stanislav Sokolov (i); 194: Steve Shoup/Shutterstock; 195: Margaret M Stewart/Shutterstock (t); Maryunin Yury Vasilevich/Shutterstock (b); 196: iStockphoto/Florin Tirlea (t); iStockphoto/Vickie Sichau (b); 197: fotohunter/Shutterstock; 198: Joy Brown/Shutterstock; 199: iStockphoto/Catherine Yeulet; 200: iStockphoto/Frank Leung; 201: Evok20/Shutterstock; 202: iStockphoto/anzeletti; 203: RAFAI FABRYKIEWICZ/Shutterstock; 204: Heather Dillon/Shutterstock; 205: Dr. Morley Read/Shutterstock; 206: mikeledray/Shutterstock (t); Alex Wild/alexanderwild.com (b); 207: irin-k/Shutterstock; 208: ajt/Shutterstock; Nh77/Shutterstock (i); 209: James Laurie/Shutterstock; 210: Brian Lasenby/Shutterstock; 211: Wilm Ihlenfeld/Shutterstock; 212: darios/Shutterstock; 213: IMAGEMORE Co., Ltd./Getty Images; 214: Oxford Scientific/Getty Images; 215: yxowert/Shutterstock; 216: iStockphoto/Paul Tessier (t); Alex Wild/alexanderwild.com (b); 217: Cathy Keifer/Shutterstock (t); draconus/Shutterstock (b); 218: Kletr/Shutterstock; 219: Joel Sartore/Getty Images; 220: Mircea BEZERGHEANU/Shutterstock; 221: iStockphoto/Tomasz Pietryszek; 222: Robert F. Sisson/National Geographic/Getty Images (t); Patti Murray/Animals Animals (b); 223: Alie Van Der Velde-baron/Dreamstime.com (l); Palto/Shutterstock (r); 224: iStockphoto/Tony Campbell; 225: DD Photography/Shutterstock (tl); Cheryl E. Davis/Shutterstock (tr); Matthew Benoit/Shutterstock (br); Karen Givens/Shutterstock (bl); 226: Chase B/Shutterstock; S.Picavet/Getty Images (i); 227: Ken Lucas/Getty Images (t); iStockphoto/gmcoop (b); 228: Graham@theGraphicZone/Shutterstock (l); Wolfgang Staib/Shutterstock (r); 229: iStockphoto/Igor Gorelchenkov; Matt Hart/Shutterstock (b); 230: sahua d/Shutterstock; iStockphoto/Andrzej Boldaniuk; KellyNelson/Shutterstock (b); 231: Dennis Donohue/Shutterstock; 232: max blain/Shutterstock; 233: iStockphoto/Floris Slooff; 234: Theodore Mattas/Shutterstock; 235: iStockphoto/mark Higgins; 236: Francois Gohier/Photo Researchers, Inc.; 237: Hedydebats/Dreamstime.com; Michael Gray/Photos.com (i); 238: iStockphoto/Jim Jurica; iStockphoto/Arpad Benedek (i); 239: Volker Steger/Photo Researchers, Inc. (t); Kim Steele/www.fotosearch.com (b); 240: Jacquelyn Lachance; 241: Damian Palus/Shutterstock; 242: Petr Nad/Shutterstock; 243: Jerzy/Shutterstock; iStockphoto/Amanda Rohde (i); 244: iStockphoto/Andrzej Boldaniuk; 245: NitroCephal/Shutterstock; 246: iStockphoto/Stephen Martin; 248: Delta Education; 249: Christian Musat/Shutterstock; 250: David Lippman/Lawrence Hall of Science; 251: Fotoline/Shutterstock; 252: Daboost/Shutterstock; 253: Jason Stitt/Shutterstock; 254: vvetc1/Shutterstock (tl); Bragin Alexey/Shutterstock (bl); Fotana/Shutterstock (t); jocic/Shutterstock (r); 255: Laurie Meyer; 256: Sandra van der Steen/Shutterstock (t); rangizzz/Shutterstock (b); 257–262: Laurie Meyer; 263: Ivancovlad/Shutterstock (l); Laurie Meyer (r); 264: Laurie Meyer; 265: Janelle Lugge/Shutterstock (t); John Quick (b).